HOUSEPLANT
Party

HOUSEPLANT
Party

FUN PROJECTS & GROWING TIPS
FOR EPIC INDOOR PLANTS

Lisa Eldred Steinkopf

COOL
SPRINGS
PRESS

Brimming with creative inspiration, how-to projects, and useful information to enrich your everyday life, Quarto Knows is a favorite destination for those pursuing their interests and passions. Visit our site and dig deeper with our books into your area of interest: Quarto Creates, Quarto Cooks, Quarto Homes, Quarto Lives, Quarto Drives, Quarto Explores, Quarto Gifts, or Quarto Kids.

First Published in 2020 by Cool Springs Press, an imprint of The Quarto Group, 100 Cummings Center, Suite 265-D, Beverly, MA 01915, USA.
T (978) 282-9590 F (978) 283-2742 QuartoKnows.com

Cool Springs Press titles are also available at discount for retail, wholesale, promotional, and bulk purchase. For details, contact the Special Sales Manager by email at specialsales@quarto.com or by mail at The Quarto Group, Attn: Special Sales Manager, 100 Cummings Center, Suite 265-D, Beverly, MA 01915, USA.

24 23 22 21 20   1 2 3 4 5

ISBN: 978-1-63159-883-8

Digital edition published in 2020
eISBN: 978-1-63159-884-5

Library of Congress Cataloging-in-Publication Data

Steinkopf, Lisa Eldred, 1966- author.
Houseplant party : fun DIY projects & growing tips for epic indoor
   plants / Lisa Eldred Steinkopf.
ISBN 9781631598838 (board) | ISBN 9781631598845 (ebook)
1. House plants--Handbooks, manuals, etc. 2. Indoor
   gardening--Handbooks, manuals, etc.
LCC SB419 .S72748 2020 | DDC 635.9/65—dc23

LCCN 2020011599

Design and page layout: Laura McFadden Design, Inc.
Photography: Heather Saunders Photography, except for page 108 and African violet images on page 103 by Chelsea Steinkopf
Illustration: Mattie Wells

Printed in China

This book is dedicated to my daughters
Hayley and Chelsea and my rock, my husband John.
Love you all!

Contents

Houseplants were all the rage in the 1970s, and then, for unknown reasons, they fell out of favor. As you presumably know (you picked up this book, after all!), their popularity has returned.

Why the resurgence? People are more aware of the healthy benefits plants have to offer. Plants provide us with oxygen, clean the air, and lower our blood pressure. Without them we wouldn't have air to breathe; they are therapeutic and can reduce stress. After a long day at work, our plants greet us when we walk in the door, and the cares of the day melt away. We may have even stopped on the way home to pick up a new friend to add to our burgeoning collection. Plants give us something to love and take care of without the intensity of a pet. The new term applied to plant owners is "plant parents." So fitting!

As a human parent myself, I understand what tremendous meaning that term has. Being a parent is serious business, so taking on the role of plant parent means making sure your plants are loved, tended to, and treated like part of the family. You're the one who will provide what they need not only to survive, but to thrive. And even better: plants don't talk back or cause any problems (for the most part, anyway). As long as you choose the best plants for your conditions (start with the cool plants I profile in the first section), meet their need for sufficient light, water them at the right time, and provide fresh air, they're going to thrive. Listen to your plants and they'll "tell" you how they're feeling. Do they seem pale and spindly? They may be telling you they need more light. Do they have leaves that are yellow or wilted? They may be receiving too much or too little water. Don't lose heart. The second section of this book tells you all about how to properly care for your indoor plants to help ensure your chances of success.

Maybe you've had a plant fatality or two and want to give up. Or you don't feel you have the fabled "green thumb" others seem to have two of. Some people may have a natural affinity for plant care, but you too can have success with plants. A "green thumb" can be yours simply by paying attention and meeting the plant's needs in a timely manner.

Once you have met your plant's basic needs, it's time to bring home some new "toys" for your green babies. It's time to spoil them! That's where the third section of this book comes in. I've put together 15 houseplant-focused projects that add to your décor and at the same time give your plants new places to hang out and grow. Some include upcycling—using things that might otherwise end up in a landfill. Some are super easy, and some take a bit more time and materials. But I promise you, they are all fun to make, alone or with a handful of houseplant-loving friends! Let's make it a houseplant party!

The Best Plants for Your Living Space

AIR PLANTS
Tillandsia species

Though they're called air plants and they do seem to literally live on air, these unique plants need more than air to survive. They don't grow in soil like most plants. Instead, they usually call some other object home, such as a tree branch or even a telephone wire. But they have essentially the same needs as terrestrial plants—light, air, and a place to anchor themselves. And perhaps most importantly, air plants need

water. If you drive through warm, tropical climates, you'll observe air plants growing just about everywhere. Did you know the Spanish moss often draped over tree branches in these places is actually a type of tillandsia?

When air plants live inside our homes, they need very specific care. First, tillandsias need bright light. A good rule of thumb is if the air plant is silver

in color, it needs more light and can be watered less often. If the air plant has green, thin leaves, it needs water more often and can be in a lower-light situation, yet a medium to bright location is still needed. These plants cannot live only on air. They need to be soaked thoroughly on a regular basis.

Misting them in between thorough soakings is beneficial, but misting should never be an air plant's primary form of receiving moisture. If your plants have good light, they should be soaked once a week. Completely submerge them in water for 30 to 60 minutes. It won't hurt them if they are in the water longer. After soaking, remove from the water, give them a good shake to remove the excess, and let them dry upside down. Species like *Tillandsia xerographica*, *T. tomentosa*, and other silver ones do not get soaked as often. I only water those every 2 to 3 weeks. After your plants are dry, return them to their growing areas.

MEDICINE PLANT, BURN PLANT
Aloe vera (*Barbadensis*)

Though it has small spines on the edges of the leaves, it's not a cactus, but rather a succulent. The foliage of aloe vera sits on top of the growing medium and produces a lot of babies called offsets. Let these baby plants grow a few inches tall, and then you can cut them away from the mother plant and pot them up individually. They are perfect for sharing with friends and make great gifts.

Aloe vera can grow leaves up to 2 feet (60 cm) tall, and if it has enough light in your home, they may send up a flower stalk of yellow tubular-shaped flowers. These plants are pollinated by hummingbirds in their natural habitats. Grow them in a well-drained potting medium, letting it almost completely dry out between waterings. (I use the word "medium" in place of "soil" as, most times, indoor potting mixes contain no actual soil. However, the words "medium" and "soil" are interchangeable.) If you repot your aloe, make sure that you plant it at the same level it was originally. Because the leaves naturally sit on top of the soil line, if they're planted too deeply, they may rot. These plants are easy to grow as long as they have bright light and are not overwatered.

Have you ever had a burn from the sun or a hot stove? You probably have. That's why keeping a burn plant, or *aloe vera*, in the house is a good idea. The sap inside the leaves soothes burns and is used in many creams and beauty products.

FAIRY WASHBOARD
Haworthiopsis limifolia (syn. Haworthia limifolia)

This diminutive native African succulent plant is one of my favorites. Unlike most other succulents, it doesn't need high light to thrive indoors. A medium to bright light is sufficient for *Haworthiopsis*. No bright south exposure? No problem. If your plant turns red instead of its normal green color, that's an indication it's receiving too much sun.

The common name of fairy washboard refers to the ridges on the leaves that resemble those of a tiny washboard. If you aren't familiar with a washboard, it was a board with ridges on it that people used to scrub their clothes on before electric washers became a thing. The fairies, I'm sure, still use these. (They don't have electricity, after all.) If you have an indoor fairy garden, this plant is a must! Fairy washboard plants are in the shape of a rosette, and as they get older, the leaves begin to grow in a slight spiral fashion. With enough light, it may send out a long flower stalk and produce small, bell-shaped, white flowers.

Use a well-drained cacti and succulent potting mix for this plant, and don't let it stand in water. Water thoroughly until water runs out of the drainage hole and then allow the potting medium to become almost completely dry before watering it again. Eventually your plant may produce offsets or babies, which can be potted up and shared with friends.

GOLDFISH PLANT
Nematanthus gregarious

This cousin of the African violet is a beautiful plant in the gesneriad family. One would never guess there are family ties as the goldfish plant has small shiny leaves that in no way resemble the fuzzy ones of an African violet. Its diminutive orange flowers resemble small puffy goldfish, thus its endearing name. The cultivar 'Tropicana', with dark-orange stripes on the flowers, is worth looking for. The natural cascading properties of the goldfish plant make it perfect for hanging baskets, which is usually how it is sold. In its natural habitat in Brazil it grows on trees, clinging to branches as an epiphyte, with no soil at all.

Full sun isn't needed, but to produce the "goldfish," it does need bright light. An east- or west-facing window is optimal. If no flowers appear, move it into more light, as light is what encourages the plant to bloom.

Keep it evenly moist to eliminate leaf drop, which may happen if it becomes too dry. Like a violet, the root system isn't large, so don't overwater. Be sure to plant it in a fast-draining potting medium containing some peat moss, such as an African violet potting mix (more on potting soils in section 2). Goldfish plants are easily propagated from stem cuttings, which can be shared with others (see section 3 on how to take plant cuttings).

LUCKY BAMBOO
Dracaena sanderiana

Though called a bamboo, this plant is not a true bamboo at all, but it does resemble bamboo with the raised rings that are prominent on its stems. Lucky bamboo originates in southeast Asia and has been used in the practice of feng shui for thousands of years. It's thought to bring good luck and happiness. Most often sold growing hydroponically (in water), it is easy to grow, adding to its popularity.

Lucky bamboo prefers medium to bright light. If grown hydroponically, change the water often. Using water without chlorine or fluoride is best, so if you have these chemicals in your water, use bottled water or rainwater. These plants can be found in different forms, including curly and braided. Buying them in

these forms is easier than trying to do it yourself, but, if you want a challenge, try it. For curly, allow the plant to grow toward the light, turning the plant gradually so it will grow toward the light in the other direction. Keep turning it until it's the shape you like. This will take patience. If you want to braid the stems, buy the stems when they are young and pliable and braid them as they grow.

You can also grow lucky bamboo in potting medium like any other houseplant. Choose a well-drained brand and make sure they never stand in water. That's ironic, isn't it? Especially after pointing out that they usually grow hydroponically. But, plants grown exclusively in water or exclusively in soil have different types of roots and thus need to be treated differently. If the tops of your stems turn yellow, cut below the yellow area, and the plant should resprout from the cut end. Don't mistake the brown parts at the base of the leaves as a problem or a "yellowing" plant. Those are old leaf sheaths that protected the new leaves as they emerged. You can easily peel those off if you find them unsightly.

MONEY TREE
Pachira aquatica

Wouldn't it be nice if a plant could grow money? You can own a money tree, but unfortunately, you shouldn't expect any money to appear. As the word *aquatica* in the botanical name implies, this plant likes water; keep it moist. In its natural habitat in Mexico, it grows in freshwater swamps and along riverbanks. While it may grow in standing water in nature, don't let it stand in water in your home, because it could end up with root rot.

Money trees may attain a height of up to 60 feet (18 m) in nature, but expect it to only reach 6 to 8 feet (about 2 m) in your home. Place this plant in bright light, such as an east or west window.

In nature, money trees produce flowers, which in turn become fruit. The fruit is eaten raw, roasted, or ground into flour. Unfortunately, however, it most likely will not produce flowers in your home.

The practice of feng shui utilizes the money tree to bring good luck. The five leaflets that form each compound leaf are thought to represent the elements: water, metal, wood, fire, and earth. It is most often sold in a braided form and can be found for purchase as a large-specimen floor plant. If your money tree develops yellow leaves, it may be suffering from low humidity. Place it on a pebble tray filled with water to raise the humidity around the plant (more on this technique in section 2). If it has brown crispy leaves, that may indicate it is too dry. Though it won't bring you monetary gain, it's a beautiful plant to grow in your home.

FRIENDSHIP PLANT
Pilea peperomioides

One of many endearing qualities of the friendship plant is the large number of small plantlets it produces. Because of this, it is easy to share with others, thus the friendship moniker. Who doesn't want to be friends with someone who shares plants? The friendship plant's perfectly round, flat leaves give it its other common name: the pancake plant. It may also be referred to as the money plant.

Pilea peperomioides is so popular, it has social media accounts dedicated exclusively to its photogenic qualities. Even better, this plant is easy to grow, as long as it has bright to medium light and is allowed to dry out a bit between waterings. Because of its succulent stem, petioles (leaf stalks), and leaves, it can be treated somewhat like you would treat a succulent plant.

The leaves of this plant like to turn quickly toward a light source. Because of this, turn the plant frequently to make sure it grows in a symmetrical fashion. This plant will produce babies on its stems, as well as in the soil around its base. If you decide to separate the babies, remove them carefully with a sharp knife. The roots are quite delicate, so cut the plantlets out of the potting medium rather than trying to pull them out. Place both types of baby plants in containers of moist potting medium. Alternatively, the plantlets on the stem can be left there, creating a multi-stemmed, fuller plant. This plant may reach 3 to 4 feet (about 1 m) tall and may require staking to remain upright.

PEACE LILY
Spathiphyllum

The white "flowers" the peace lily sends up are one of the reasons for its popularity. This plant does not need full sun to bloom, nor does it want to be in that much light. Instead it prefers a medium light. It may bloom continually when in the correct lighting situation.

You might assume the elliptical-shaped appendage cupped around the knobby spike is a flower, but it's not. It's a spathe, and the actual flower is the knobby spike in the center (called a spadix). Often you will see white dust on the leaves of your plant. You may panic, thinking there are bugs or a disease present, but don't worry. The spadix produces white pollen, and sometimes it falls on the leaves. If that dusty look bothers you, wipe it off or leave the spathe and cut the spadix off. The spathe will give you the look of a flower, and the messy pollen-maker will be gone.

Peace lilies let you know when they need water by wilting over the side of the pot. The good thing is, unlike many other plants, it will perk back up soon after you give it water. Do not use this as an indicator of when to water, though. It works, but the consequences will be yellow leaves and brown tips if it happens too often. Instead, always keep your peace lily moist, but not standing in water.

PONYTAIL PALM, ELEPHANT FOOT, BOTTLE PONYTAIL

Beaucarnea recurvata

It's a good thing we have Latin scientific names for plants, because they can have many common names, as is the case with *Beaucarnea recurvata*. Most often, though, it is known as the ponytail palm—even though it's not a palm at all. If you've ever seen one, it will be obvious why the ponytail name was chosen. The long, strappy leaves start out very tight together where they come out of the stem. They look as if they are tied together with an elastic band.

The swollen bulbous stem (called a caudex) resembles the leg of an elephant, especially as the plant ages. This large stem stores water like a camel's hump, and because of this feature, this plant doesn't need to be watered often. In fact, a ponytail palm should not be watered until the potting medium is almost completely dry. It is a light-loving plant originating in parts of the southwestern United States. If it's placed in a low-light situation, the usually thick, leathery leaves may become thin and floppy.

In its native habitat, a ponytail palm can reach up to 30 feet (9 m) tall! But inside, it will only reach a few feet (less than 1 m) in height, and this over many years as it's a slow grower. Use a clay pot when planting ponytail palms; it allows evaporation of water through the walls of the pot, helping prevent the soil from becoming waterlogged.

POTHOS, DEVIL'S IVY
Epipremnum aureum

Are you trying to attain the "jungle look" with vines scrambling up your walls and around the windows? If so, pothos is the plant for you! This iconic houseplant has been climbing walls and framing windows for decades. It can live in full sun, yet also will do well in a low-light situation. In our homes it will remain in its juvenile form, never producing the enormous leaves with splits in them, not unlike their relative, the monstera plant. Most often, the plant you purchase will have yellow markings on the leaves. If it's placed in too little light, the yellow markings will disappear on the new growth. If that happens, place it in a location with higher light, which will help new growth again display the yellow markings.

Pothos plants also come in bright green, dark green, and variegated white forms, so there are many to choose from. If you have a low-light situation, choose the non-variegated, dark-green form; the variegated forms need more light.

Keep this plant evenly moist. It is forgiving of occasional dry conditions but will develop yellowing leaves and may drop its oldest leaves, leaving bare stems behind. It is easy to keep the plant full by trimming the long stems back to the soil level. New shoots will arise from the cut stem, creating a fuller plant.

Use a 4- to 6-inch (10- to 15-cm) piece cut off the end of the stem to propagate a new plant. This cutting can be planted on its own, planted back into the same container to keep the plant looking full, or shared with a friend (see section 3 for how to propagate plants via cuttings).

SHREK'S EARS, ET'S FINGERS, HOBBIT'S PIPE, TRUMPET JADE

Crassula ovata 'Gollum'

This easy succulent will be a conversation piece in your plant collection. Depending on which movie you've seen, you may prefer to call it Shrek's ears or ET's fingers. It's also said that the name Gollum came from Tolkien's character from *The Lord of the Rings* trilogy. Regardless of what you call it, this plant is a monstrose form of the original jade plant (*Crassula ovata*), which has flat leaves. Monstrose is the term used for funky forms of regular plants that come about through chance genetic mutations.

Shrek's ears is an easy-to-care-for succulent, as long as it gets enough light. A bright south- or west-facing window is preferred. Because of its water-storing capabilities, it can survive for quite a while without water. If the leaves start to pucker, it needs water. Water thoroughly at that time but try not to let it get that dry again. Plant Shrek's ears in a cactus/succulent potting mix to provide good drainage. Clay pots are a great choice for succulent plants as they are porous and excess water can evaporate out the sides of the pot. This plant could eventually grow up to 3 feet (91 cm) tall and 2 feet (61 cm) wide.

SNAKE PLANT

Sansevieria trifasciata 'Bantel's Sensation'

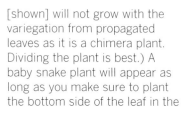

It may surprise you to know that the snake plant has received a lot of bad press in the past. Surprising because this plant has become a collectible, with new cultivars being created all the time. It also has the common name mother-in-law's tongue because of the sharp edges of its leaves, which doesn't paint mothers-in-law in a good light. The snake plant is commonly used as a low-light plant, though it can often be found perishing in dark corners. While these African succulents will survive in low light for a few years, they much prefer high light. In enough light they may even send out bloom stalks, and their strong underground stems (rhizomes) may grow so zealously that they break the container they are growing in!

Snake plants are thick-leaved succulents. They're quite drought tolerant and don't want to be kept in a wet potting medium. If you feel it may need moisture but are not sure, err on the dry side to be safe.

Propagate your snake plant by cutting a single leaf in 2-inch (5-cm) increments and planting each piece in moist potting soil, first allowing the cut end(s) to callus over for a few days. (Please note: Bantel's Sensation [shown] will not grow with the variegation from propagated leaves as it is a chimera plant. Dividing the plant is best.) A baby snake plant will appear as long as you make sure to plant the bottom side of the leaf in the medium. Your snake plant can also be propagated from offsets, separating the baby plants that will rise from the potting medium. You will need a knife to cut through the rhizomes.

SPIDER PLANT, AIRPLANE PLANT
Chlorophytum comosum

Many people have sentimental feelings toward the ubiquitous spider plant. Perhaps your mother or grandmother had one in her window. Some describe the dangling plantlets as "spiders," while others call them "airplanes." Both evoke pictures of plants in the air, whether soaring like planes or swinging from silken threads like a spider. However they are described, the appendages are captivating, adding movement to the plant. They are exceptional hanging plants, but also make a great "hairdo" for a face-shaped pot.

There are variegated and plain green forms, and both need medium to bright light or the leaves will stretch, become weak, and flop over. If they receive enough light, they will produce diminutive white flowers and multiple runners with baby plantlets.

Keep your spider plant evenly moist and the humidity high to prevent brown tips on the leaves. Fertilizer salt buildup or fluoride in municipal water sources can also cause brown leaf tips. Trim the brown ends when they develop, and remedy the situation causing them.

Spider plants have fleshy, tuberous roots that quickly fill the pot.

They may distort a plastic pot or even push the plant up out of the container. When this happens, up-pot your plant to the next size container.

To propagate your spider plant, remove the baby plantlets and place them in soil or water, roots down. You may also leave the plantlet on the runner attached to the mother plant, pin it to another container of moist soil with a piece of bent wire or a paper clip, and allow it to root while still receiving water and nutrients from its mom. After it begins to grow, cut it off the stem that attached it to its mother.

SWISS CHEESE PLANT, FRUIT SALAD PLANT
Monstera deliciosa

Imagine a plant so popular it has taken over social media one day a week. This is the plant that #MonsteraMonday is all about! Its lesser-known name of fruit salad plant refers to its fruit, which tastes like a cross between a pineapple and a banana. The unripe fruit is not edible, though, and may cause mouth and throat irritation if ingested.

This large plant was a fixture of mid-century modern decorating, and the lofty ceilings and open floor plans popular today have allowed this plant to come back into vogue. The enormous fenestrated (hole-filled) leaves are unique, and the holes help the plants cope with the strong winds and large amounts of rain they receive in the jungles they call home. You may be confused when you buy your plant, because its juvenile form doesn't always have split leaves, but they'll appear as the plant ages.

For additional moisture, monsteras send out aerial roots that also help stabilize the plant. In our homes, these roots attach themselves to floors or walls and leave marks when removed, so watch for them and cut them off, or place them back into the container. As your monstera grows in size, you may find a trellis will be helpful to support your plant (see section 3 for a DIY wooden plant trellis project).

Monstera can tolerate low light but prefers medium to bright light. In their native habitat, they start life on the jungle floor and scramble along until they find a tree to cling to, climbing to the top for light.

Keep this plant evenly moist, letting it get quite dry before watering again. All you need to propagate a monstera is a leaf attached to a small piece of the stem. It can be rooted in water or in a container filled with moist potting medium.

VENUS FLYTRAP
Dionaea muscipula

These carnivorous plants have specialized parts that attract insects, and after capturing them, the plant digests the insect and uses the nutrients they provide. Why do they need to eat insects, you ask? They live in such poor soil conditions, they need to find nutrients another way.

The modified leaves of the Venus flytrap are like a large mouth with hairs on the edges. The mouth is often red inside and contains nectar glands that attract certain insects. There are small trigger hairs on the inside of the trap, and when an insect lands inside and touches the same hair twice within 20 seconds, the leaf snaps shut. The hairs on the edge of the mouth become like the bars of a prison, keeping the insect in. Immediately, the plant releases enzymes that begin to digest its prisoner.

Venus flytraps need bright light, high humidity, and a highly acidic potting mix containing mostly peat moss. Yes, they make their own food by means of photosynthesis, but the nutrients the insects provide help the plant grow better. If you can direct some flies their way, it's helpful, but the plant will survive without them. Contrary to popular belief, Venus flytraps shouldn't be fed hamburger or other raw meats because the plant does not release the digestive enzymes unless the food source is moving around.

These plants go dormant for about 5 months during the winter season. Throughout this natural dormancy period, they need to be in a cold place, just above freezing. A garage or an unheated room works perfectly. With a little bit of extra care and some insects to eat, these unusual plants will thrive.

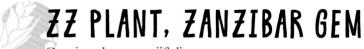

ZZ PLANT, ZANZIBAR GEM
Zamioculcas zamiifolia

rhizomes develop leaves on top and roots on the bottom.

An unusual attribute of this plant is that new plants can be grown from an individual leaflet. However, patience is needed. Just place one leaflet in moist potting medium or water. It can also be propagated from an entire stem or divided by cutting through the rhizomes to separate smaller plants from the larger one.

Though the ZZ can do well in low light, it prefers medium to bright light and can become quite a large plant.

If you have a low-light area and everything you've tried growing dies, you haven't tried the ZZ plant! This East African native plant sports glossy, dark-green leaves that make a strong architectural statement and lend a modern vibe to any room. The leaves are upright and made up of many leaflets on each rachis (stem of a compound leaf), the actual "stem" being the underground tuberous rhizomes, not unlike a snake plant. The

Because of its large fleshy rhizomes, this plant is drought tolerant and doesn't appreciate an overly wet potting soil. If it's in a higher light situation, it will need more water, and if it becomes too dry, it will drop leaflets.

Taking Care of your Houseplants

Like a dog, cat, or any other pet, houseplants need your care to survive and thrive. Of course, they don't need to go for a walk and there won't be any poop to scoop, but there are still things that a plant needs. These include water, light, food, shelter, and of course, love. We're going to talk about a few important factors to consider when it comes to caring for your houseplants.

Light

Though you need to put food out for your pets, plants make their own food through photosynthesis. Without going into great detail (this isn't science class, after all!), let me explain a little about the process. When light falls on the green leaves of your houseplants, the chlorophyll inside the cells collects that light. Along with the light, the plant uses water (which you provide), carbon dioxide from the air, and minerals to make glucose/sugar, which is the food that fuels the plant's growth. A by-product of this process is oxygen, which, of course, we humans need to live. As it photosynthesizes, the plant is oxygenating and cleaning the air at the same time.

In order for photosynthesis to occur, plants need light. The amount of light they need depends on the conditions under which they naturally grow. How do you know if you have enough light to keep a plant alive? If you can read in your room during the day without a light on, you have enough light to support a living plant. (No faux plants, please.) There are different levels of light, however, and you need to determine if you have high, medium, or low light before choosing which plant to add to your home.

To determine the level of light you have, consider which direction your window faces.

- If the sun comes up in your window (probably earlier than you would like it to), it faces east. East light is a medium to bright light and is a soft, low-temperature light.
- If the sun goes down in your window, it faces west. This is also a medium to bright light, but it is a hotter light because it comes in the window in the afternoon.
- If you have light in your window almost all day, it faces south, and that is a great window for cacti and other succulents, as well as other high-light plants.
- If the sun never shines directly in your window, it faces north. This window is perfect for low-light plants, such as ferns and Chinese evergreens.

Determining the amount of light your window receives is key to deciding which plant to bring home. If a plant doesn't have enough light to survive and grow, it will slowly deteriorate and eventually die. Choose the right plant for the light you have, and you'll be the owner of a thriving plant!

Temperature

Most of our indoor plants come from warm climates. Different plants like different temperatures, but a good rule of thumb is: if you are comfortable, your plant probably is as well. If, on the other hand, you like to keep your home below 55°F (13°C), that might not work for your plants. Many houseplants prefer the temperature around them above 55°F (13°C), and warmer would be better. The range between 55°F and 85°F (13°C and 29°C) is acceptable for most plants. Temperatures higher than 80°F (27°C) may cause damage to some plants, especially if they are dry. They can take higher temperatures if the humidity is elevated, because they are not losing water as fast through transpiration. If the temperature falls below 55°F (13°C), some plants may suffer cold damage. Although temperature requirements differ from plant to plant, all plants like a slight temperature drop at night. This mimics the nighttime fall of temperature in nature.

Water

There is much discussion and controversy when it comes to watering houseplants. I'm going to tell you what works for me and will work for you. First, all plants are watered in the same way: Add water until it runs steadily out of the drainage hole. This includes cacti and other succulents. The difference is the amount of time between waterings. For plants that like it on the drier side, water less frequently. For plants that prefer more moisture, water more frequently. By watering so the water runs completely through the container, you are wetting the entire root ball, as well as drawing air down through the root area.

How to water a plant in a pot with a drainage hole:

A pot with a drainage hole is important, especially if you're a new houseplant parent. If there's no drainage hole, it isn't easy to determine whether or not a houseplant has enough water. After all, you won't know if the plant is standing in water or if the root ball was moistened all the way to the bottom. If you have a drainage hole, carry the pot to a sink or bathtub and pour water over the soil. Giving your plant a shower at the same time is a great way to clean the leaves. Keep adding water until it freely drains out the drainage hole. Let the pot sit in the sink or tub until it finishes draining, then move it back to its home base.

How to water a plant in a decorative container:

When you water a plant whose pot is nestled inside a more decorative container (cachepot), take it out of the decorative container completely, water it, let it drain fully, and then return it to the decorative container. This method ensures that your plant is never standing in water, which will lead to root rot.

How to water a plant with a saucer beneath the pot:

When you water a plant in a pot with a saucer, the excess water will drain into the saucer instead of draining out of the hole and down the sink. If water is standing in the saucer 30 minutes to an hour after watering, empty the saucer into the sink. This may be difficult if the container is large and hard to move, but the roots of your plant may rot if water is left standing in the saucer for long periods of time. A turkey baster is a good tool to include with your houseplant tools; use it to remove the excess water from the saucer.

NO DRAINAGE HOLE? NO PROBLEM!

If you find a container you must have and it doesn't have a drainage hole, there are a couple of ways to handle it. If you have a drill, you can purchase a masonry or diamond-tipped drill bit and carefully drill a hole in your container. If you cannot drill a hole, I suggest you use your container without a drainage hole as a cachepot (cachepot means "hide a pot" in French). In other words, your houseplant will stay in the grower's pot it came in, and you will simply slip the plant into the decorative container so the grower's pot will be hidden.

layers of soil may have a different moisture level than the soil at the top. In this case, use a large dowel or bamboo stake, sticking it all the way to the bottom of the pot and leaving it there for a few minutes. When you pull it out, check the bottom of the stick, and if it's wet, don't water the plant. See photos below.

Fertilizer

Did you know that fertilizer is not food for plants? The only food a plant receives is from the light

How do you know when to water again?

The first thing to know is to never water a plant on a set schedule. Many factors must be taken into account, including the amount of sunshine and temperature fluctuation. A good idea is to *check* your plant on a schedule to determine if it needs water.

There are many different water meters you can buy to check for moisture, but the best way is to stick your finger in the soil. If it's dry up to your second knuckle and the plant prefers to be consistently moist, it's probably time to water. If you have a large floor plant, using your finger isn't the best way to check the moisture level. The bottom that falls on it. Fertilizing is something you don't have to do—your plant will do just fine with sunshine and nutrients from its potting mix. However, it is beneficial to fertilize a plant, as long as you don't overdo it. Fertilizing a plant is equivalent to humans taking vitamins. It provides the plant with extra nutrients.

Fertilizing should only be done when the plant is actively growing. If you live in an area where daylight hours are shorter in the winter season, your plants won't need fertilizer from fall until spring. When you see signs of new growth in the spring, start your fertilizing regimen again. I fertilize my plants every fourth watering, or approximately once a month, using a water-soluble fertilizer with even numbers, such as 10-10-10 or 20-20-20.

If you use an organic fertilizer, such as fish emulsion, those three numbers will be much lower and may not be equal, which is perfectly natural. Some people prefer to use only organic fertilizers for their plants. One saying that definitely applies when fertilizing plants is "less is more." I never use the fertilizer at full strength or use more than the directions call for. If you choose, you could use a small amount of fertilizer, such as one-fourth strength, in your water every time you water your plants.

THE MEANING OF FERTILIZER NUMBERS

The three numbers on a fertilizer container represent, in order, the percentage of nitrogen, phosphorous, and potassium found in the fertilizer. A good way to remember those elements and how they help your plant is the old saying, "Up, down, all around."

- The first number represents nitrogen, which helps the "up" part (the green part) of the plant.

- The second number represents phosphorous, which helps the "down" part of the plant, meaning it helps promote strong root growth. It also helps the flowers grow larger, last longer, and have a richer color.

- The third number represents potassium, which helps with the "all around" health of the plant.

QUICK TIPS FOR HOUSEPLANT CARE

How to improve your light:
To help maximize the amount of light reaching your plants, wash your plants to remove dust and grime blocking their photosynthetic process. Wash your windows, especially when bringing your plants in after a summer sojourn outside.

Choose a light paint color on your walls to reflect light to your plants. Use mirrors in strategic places to bounce light around the room and give your plants a boost.

How to raise the humidity:
Plants love humidity. Most houseplants are found growing naturally in humid climates, such as rainforests. Use a room humidifier, or if that isn't feasible, a pebble tray. Place your plant pot with its saucer on a larger saucer or tray filled with pebbles and water. Keep the pebble tray filled with water, and as it evaporates around your plant, the humidity will rise. If you choose not to use a saucer under your pot, make sure the pot isn't sitting directly in the water; it should be propped up by the rocks. If misting is something you like to do, go for it. Just know that it isn't really doing much for your plant.

How to improve flowering:

Fertilizer does not make a plant flower. If you have a flowering plant that is not flowering at some time in a yearly cycle, it likely needs more light, not fertilizer. Move it closer to a light source or add an electric light if it isn't possible to move it to another location.

How to groom plants:

When a plant leaf has turned yellow, it won't turn green again. If it has a brown tip, you can trim the tip off, cutting it the same shape as the original leaf. If most of the leaf is yellow, it's time to cut the leaf off entirely. Cut it back to the stem it originates from.

How NOT to improve drainage:

Many people put gravel or pot shards (broken pieces of pots) in the bottom of a container because they believe it helps with drainage. It doesn't help and can actually hinder the drainage process. I use a piece of window screening to cover the hole, allowing water to escape easily while the potting medium stays in.

15 Houseplant Projects and Crafts

MARIMO BALL AQUATIC GARDEN

Quite often plants die because they are overwatered. The best thing about aquatic plants is that you can't overwater them because they **live** in water! A marimo ball (*Aegagropila linnaei*)—also called a moss ball, lake ball, or lake goblin—is a perfect plant for your desk or bedside table, as long as it receives medium light.

What is a marimo moss ball, you might wonder? The first thing to know is that it isn't moss at all, but a type of algae. Their native habitat is freshwater lakes in Japan and Iceland, where they thrive rolling back and forth on the lake bottom with the waves. This gives them their spherical shape. In Japanese, *marimo* means "bouncy ball."

As they live at the bottom of lakes, marimo balls can grow in low to medium light, which makes them perfect for home environments. If they're exposed to too much bright light, they will suffer and turn brown. In addition, unwanted algae will grow if the container receives too much light, which may also block the amount of light getting to the marimo ball. Marimo balls need to be turned regularly to expose all sides to the light, otherwise brown, dead patches develop. Mimicking the movement of their natural environment helps keep their round shape.

MATERIALS

A glass display container (Choose a larger container if you'd like to add fish to your aquatic garden—see page 43.)

Water (Tap water is fine.)

Marimo moss ball(s)

Gravel or sand for the bottom of the container (optional, unless adding other plants with roots)

Decorations (optional)

Optional additional supplies:

Betta fish

Water conditioner (If you add a fish to your container, you must use a water conditioner to remove the harmful chemicals from the water, such as chlorine and fluoride.)

Betta food (They're carnivores and don't eat plants.)

Cryptocoryne plant

STEP 1 Make sure your glass container is clean. If you wash it or use window cleaner, make sure it's well rinsed to get all the cleaning product residue removed.

STEP 2 Add gravel to the bottom of the glass display container.

STEP 3 Fill the container with water.

CARING FOR YOUR MARIMO BALL AQUATIC GARDEN

Marimo balls can grow up to 12 inches (30 cm) around, but in an aquarium, they may grow to only 2 to 3 inches (5 to 8 cm). They grow very slowly and are long-lived plants that require minimal care.

- Change the water occasionally and rinse the marimo ball to make sure no algae or dirt is collecting in the moss ball, especially if there are fish in the tank.

- If the water starts growing algae because it receives too much light, change the water, rinse the ball more often, and move your aquatic garden to lower light.

- Swish the bowl around occasionally so your moss ball keeps its shape and all sides receive light.

- Marimo balls don't need any fertilizer because they get their food from the sun. Adding fertilizer to the water might trigger growth of unwanted algae.

4

AN EVEN BETTA ARRANGEMENT

If you have a container that holds 2½ gallons (9.5 L) of water or more, you can add a betta fish (*Betta splendens*), otherwise known as a Siamese fighting fish, for more interest. You should only have one fish at a time in your tank, as they will fight. If you want to add other plants in with the marimo that are compatible with these fish, look for cryptocoryne (commonly called "crypt"), which is readily found at big box pet stores or aquarium stores. Like the marimo, it's an easy, low-maintenance aquatic houseplant.

If you add a betta to your container, some of the water will have to be replaced every week. Betta fish bring added responsibility. They must be fed regularly, and the tank needs to be cleaned more often. But they also bring beauty and movement to your new water world. You decide what is best for you.

STEP 4 Add your marimo balls to the container. They may float, so squeeze them gently to remove any air bubbles, and they should sink to the bottom.

STEP 5 You can add a cover or leave it open. A glass plate makes a great cover and doesn't block the light.

Enjoy your marimo balls for years to come!

RECYCLED T-SHIRT PLANT HANGER

Think twice before you throw away that old t-shirt. It can be upcycled into a plant hanger in just a few simple steps. Maybe you've admired those macramé hangers that are all the rage and can make one from the pattern on page 48.

If that seems too time-consuming, grab a t-shirt, some scissors, and a tape measure, and you can have a great plant hanger in a short amount of time. It only entails cutting up the shirt and tying 10 knots.

You could make a few of these in an afternoon and upcycle old clothes at the same time. Before you know it, your windows will be full of hanging plants!

MATERIALS

Large (or larger) t-shirt (You can use a smaller size but will have a smaller hanger.)

Scissors (Make sure they can cut fabric easily.)

Tape measure or ruler

Pencil or pen to mark the cutting lines

Potted plant in a 4" to 6" (10 to 15 cm) container

Ceiling hook to hold the plant hanger

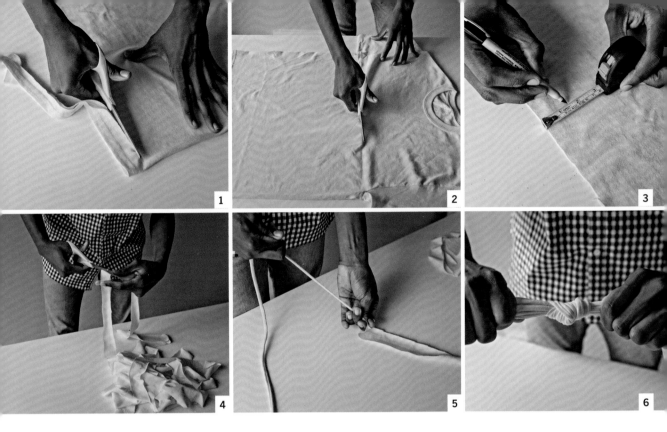

STEP 1 Lay the t-shirt on a flat surface and smooth it out to remove any large wrinkles. Cut the hem off the bottom of the shirt.

STEP 2 Cut the t-shirt across the front directly under the sleeves and discard or save to use for another project. You now have the body of the t-shirt left.

STEP 3 Using a tape measure or ruler, measure up 1½" (3.5 cm) from the bottom of the shirt. Mark the spot on both sides. You may want to use a yardstick to draw a straight line from one mark to the mark on the opposite side. It doesn't matter if it isn't perfect. Cut the body of the t-shirt into eight strips 1½" (3.5 cm) wide starting on the side edge of the shirt.

STEP 4 These pieces will be circles when first cut. To make them into strips, cut through the circle in one spot.

STEP 5 Pull the strips to stretch them out. When stretched, the entire strip will curl up like a tube.

STEP 6 Gather all the strips together, making sure they are even at one end. The hanger is worked from the bottom to the top. Holding all the strips together, tie a large knot approximately 5" (13 cm) from the bottom.

STEP 7 Divide the eight strips into pairs, forming four pairs of two strips each. Using one pair of strips, tie a knot approximately 3" (7.5 cm) from the large bottom knot. Continue around, tying a knot in each set of strips. There will be four knots.

STEP 8 Now drop down another 3" (7.5 cm), take one strip from one pair and one from the adjoining pair, and tie a knot.

STEP 9 Continue around, tying three more knots. This forms the basket that the plant will be placed in.

STEP 10 Now gather all eight strips back together and tie a large knot at the top of the hanger. Cut the ends off above the knot. You can adjust the length of your hanger by tying the knot lower toward the basket and cutting the long ends off, or you can leave them hanging.

HOW TO HANG YOUR T-SHIRT PLANT HANGER

Place a hook in the ceiling where you'd like to place your hanger. Make sure it's in a spot that has an appropriate amount of light for the plant you are using. Nestle the plant into the fabric basket and hang it on the hook.

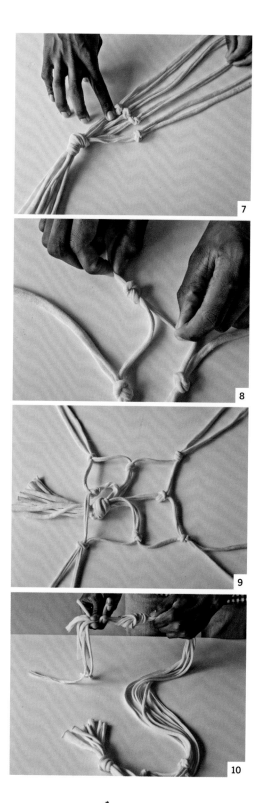

MACRAMÉ PLANT HANGER

Are you running out of floor and windowsill space for all your plants? Look up! By making a plant hanger, you can utilize the ceiling area to house your vining plants. Macramé has made a comeback, and it's even more popular now than it was in

the 1970s. The technique simply involves tying different knots with string to form a pattern. If you have ever made a friendship bracelet, you can macramé, because they use many of the same knotting techniques. Making a hanger for your

plant is fun and easy once you master a couple of those simple knots. The best part is, you can have more plants by hanging them from the ceiling, making the most of unused space. You only need a few supplies to make this hanger.

Note: *Follow the directions exactly as they are written, one step at a time, and you macramé hanger will be simple to make. Pay attention to the repeats.*

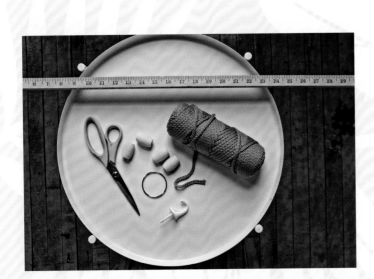

MATERIALS

34 yards (31 m) of 4 mm macramé cord in your choice of color

4 20 to 25 mm large-hole macramé beads in a round or oblong shape

1" to 2" (2.5 to 5 cm) metal or plastic ring

Scissors

Measuring tape or yard stick

Lighter or match (optional)

Ceiling hook specifically for hanging plants

KNOTS USED

square knot

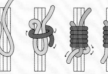

wrap (or gathering) knot

half knot

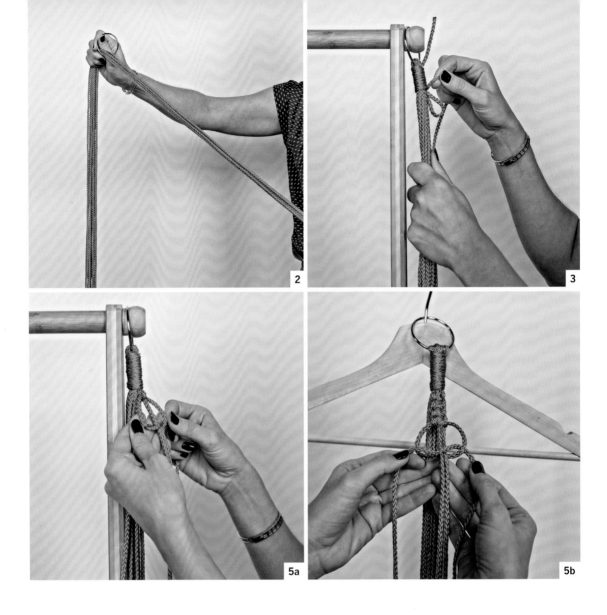

STEP 1 Cut eight 4-yard (3.7 m)–long cords. Cut two 1-yard (91.4 cm)–long cord for wraps.

STEP 2 Thread the eight 4-yard (3.7 m) cords through the ring so that the ring is at the middle of the cords. There will be sixteen cord ends hanging down from the ring. Make sure they are relatively even at the bottom. This hanger is worked from the top to the bottom.

STEP 3 Tie a 1½" (3.5 cm) wrap (or gathering) knot (see knot illustrations on page 49) around the sixteen cords using one of the 1-yard (91.4 cm) cords.

STEP 4 Now you will begin to work on one arm or sinnet of the plant hanger. Each sinnet is worked separately until the basket to hold the plant pot is formed. Separate the sixteen cords into four groups of four strings each to form a sinnet.

STEP 5 Using four cords, tie five square knots (see knot illustrations on page 49).

STEP 6 Switch threads, bringing the two longer middle threads to the outside. (The longer cords will now be used to tie the knots.) Drop down 2" (5 cm). Tie ten half knots (see knot illustrations on page 49). The result is a twisted stitch.

STEP 7 Thread one bead over all four cords and slide it up right under the half knots you just tied. Tie ten half knots under the bead.

STEP 8 Drop down 6" (15 cm) and tie two square knots. Repeat the process from Step 5 through Step 8 on each of the remaining three sinnets. The four arms or sinnets of the macramé hanger are now done, and we will start to work on the basket of the macramé holder that will cradle the plant container.

A TRICK FOR STOPPING THE FRAY

If you would like to stop the cord from fraying, tie a knot at the bottom of each cord. Or if the cord used is heat fusible, burn the ends with a lighter or match to make sure they do not fray. This is done by quickly passing the flame over the end of each cord. It should be done carefully so the flame or the melted material doesn't burn you. A knot tied at the end of each heat-fusible cord also keeps it from fraying and adds a different look to the hanger. *Safety note:* This will **not** work with cotton cord.

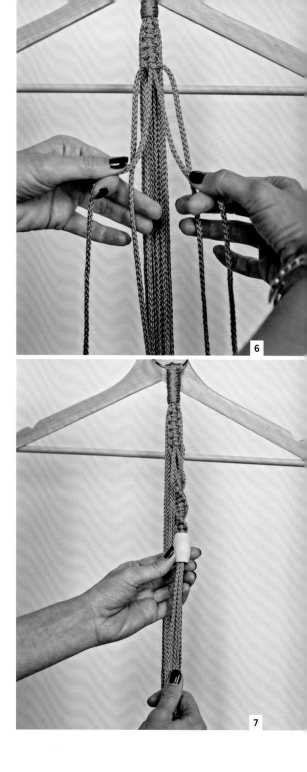

A HELPFUL HINT

While I'm working on my macramé, I hang the ring from a wreath hanger that is over a door, but any over-the-door hook would work. You could also use a nail on the wall or hang the ring over the top of a clothes hanger hung from a nail or curtain rod. My friend uses a doorknob on her kitchen cupboard. The point is, it is easier to work on if it's hung from something. Do whatever works best for you and use what you have available.

Forming the basket:

STEP 9 By using threads from neighboring sinnets each time, the result will be a webbed basket that the plant pot sits in. Skip down 4" (10 cm). Take two cords from one sinnet and two cords from the neighboring sinnet and use them to tie two square knots. Repeat with the remaining sinnets.

STEP 10 Once all four sinnets are tied, repeat the step above, again skipping down 4" (10 cm) before tying the knots.

STEP 11 Drop down 2" (5 cm) below the last square knots made and tie a wrap knot with the remaining 1-yard (91.4 cm) cord. Trim the ends of the cords evenly and let the ends fray naturally.

9

10

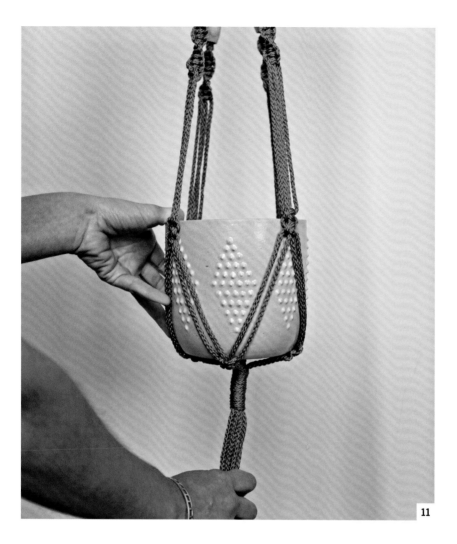

11

HOW TO HANG YOUR MACRAMÉ MASTERPIECE

Your macramé plant hanger is finished and ready to be hung from the ceiling. I suggest buying a specialized plant hook that is meant to bear the weight of the container, plant, and moist potting medium. Finding your plant on the floor and the pot smashed because you used a hook that was too small is heartbreaking. The amount of weight a hook will hold will be displayed on the packaging. Make a few hangers to fill your window with greenery. Because they hang from the ceiling, it leaves plenty of room on the floor and windowsill for more plants. Time to go shopping!

A HAWORTHIOPSIS ZEN GARDEN

We all need to shut down, turn off social media, and contemplate life from time to time. As we know, that can be hard to do. This desk-sized Zen garden is a perfect way to do just that. What is a Zen garden, you ask? It is a small Japanese rock garden or "dry landscape" garden. It is usually an outside garden, with purposefully arranged rocks, water features, plants, and gravel or sand to rake.

Many of these gardens are found at temples in Japan, where the priests often rake the gravel to help with concentration. The raking of the sand in a ripple pattern represents water. Rocks are usually a part of the garden, used to represent mountains, but because of the small scale of our garden, they haven't been included. If you would like to add a small stone or two to make it more Zen-like, feel free to do so. This is your creation and there are no rules that must be followed.

Though this isn't a true Zen garden, it does give us the same vibe. Place it on your desk, and when you are feeling stressed, pick up the rake, admire the plants, and let calmness enter your mind.

The plants we used are succulents that can withstand lower light levels than most. If your garden is in a medium to bright light, it should grow well.

MATERIALS

Two small haworthiopsis (formerly haworthia) plants (For more on these plants, see page 14.)

Decorative container approximately 6" to 10" (15 to 25 cm) wide (with a drainage hole). A shallow pot is best.

Sand (colored or plain)

Mini rake

Potting medium (A well-drained cactus and succulent potting medium is best.)

1

2

3a

STEP 1 Add the cactus and succulent potting mix to your container. Leave approximately ½" to 1" (1 to 2.5 cm) of space at the top of the container for the sand. If the potting mix is too close to the top, the sand will wash over the side of the container when you're watering the plants.

STEP 2 Plant your haworthiopsis along one edge of the container. Carefully snuggle the plants in close to each other so you have more room for the sand area of the design. Lightly firm the soil around the plants and across the top of the container to make a level plane for the sand. Water the plants to settle them in.

STEP 3 Now it's time to add the sand. I used two colors of sand, but plain-colored sand can also be used and is more in keeping with the Zen garden aesthetic. Use a funnel or small container to add the sand, being careful to keep the sand off the leaves of the plants, especially if colored sand is used.

STEP 4 If you do get some sand on the leaves, use a small brush, such as a paint or makeup brush, to carefully sweep it off.

STEP 5 Place your Zen garden on your desk or somewhere you'll be able to use it often to settle your thoughts and bring calmness to your day.

WATERING TIP

Be careful when watering your Zen garden to make sure the potting medium and sand aren't mixed together. If that occurs, just add more sand to cover the area again. For more on growing haworthiopsis, see page 14.

TEST TUBE PROPAGATOR

If your plants become long and straggly or look like they're in need of a trim, you now have a great reason to start new plants! Trimming your plants helps them stay full and keeps them at a manageable size. When you trim a long stem off a vining plant—such as a pothos—a new stem will emerge near the cut area. The cuttings can be rooted directly in moist potting medium or in water.

Rooting the stem pieces in water only requires a receptacle that holds water. You can use old spice bottles, small vases, or even drinking cups to root new plants, but why not use something that also doubles as artwork? These test tube propagators are a perfect fit. The tubes are attached to small pieces of wood and then hung on the wall. One is perfect for rooting a single cutting, but a whole wall of them can make a bold statement and hold multiple cuttings, creating a green wall. These propagators are simple and quick to make, and buying the supplies won't break the bank.

MATERIALS

Test tube

Wood rounds that are approximately 2" to 3" (5 to 7.5 cm) around (I used birch wood found at my local craft store.)

Sawtooth picture hangers 1¾" (4.5 cm) wide

Clear bumper pads

Industrial-strength glue

Rubber bands

Hammer

Nail or screw to hang it on the wall (A removable adhesive hook would also work.)

Water

Plant cuttings to be propagated

1

3a

3b

STEP 1 Gather your materials. Because you'll be using glue, cover your work area with newspaper. Decide which side of the wood round you like best—this will be the front side, where you'll attach the test tube. Turn it over and nail the sawtooth picture hanger at the top of the round. Add a bumper to the bottom of the round below the hanger so it will hang level on the wall.

STEP 2 Now turn the round over. **IMPORTANT:** Make sure you have the sawtooth hanger level at the top before your glue your test tube on. If you don't check, you may have a test tube that doesn't hang straight. Imagine the saw-tooth hanger on the back is the top of a T and the test tube is the leg of the T. Hold the test tube on the front of the wood round, and at the same time turn it to look at the back to make sure the test tube is perpendicular to the hanger. You can place the test tube top edge even with the wood round edge or raise it up an inch or so, or you can center the wood round in the center of the test tube. The placement of the tube on the wood round is up to you.

STEP 3 Carefully lay it down, remove the test tube, and run a generous line of glue down the center of the wood round. Place the test tube in the glue, pressing it firmly but carefully to make sure it has maximum contact with the glue. Don't press too hard, or you may break the glass. **FYI:** Do *not* use a hot glue gun. I tried it, and it worked for about an hour before the test tube fell off the wood and smashed into a million pieces on the floor.

STEP 4 Carefully stretch the rubber bands around the test tube and wood piece to keep the test tube in place while the glue dries. Allow it to dry for a few hours (follow the directions on the glue container) before taking the rubber bands off.

STEP 5 Add water and plant cuttings to the test tube, and you are ready to hang your propagator on the wall. Remember to keep adding water as needed, especially as the roots start to grow and use water.

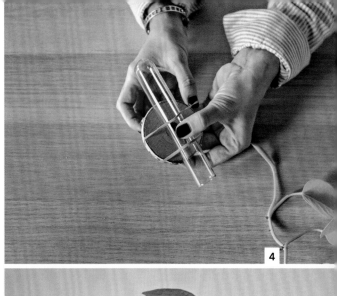

PLANTS TO TRY

- Pothos
- Heart leaf philodendron
- English ivy
- Tradescantia
- Lucky bamboo
- Trailing peperomias

- Purple passion vine
- Spider plant
- Pellionia
- Aeschynanthus
- Cissus varieties such as grape ivy

You don't necessarily have to use a trailing plant. ZZ plant and aglaonema also work well. Make one or as many as you want. They add a lot of beauty to a wall while your plants are busy growing roots.

RECYCLED TABLE KNIVES PLANT STAKES

Do you name your plants? Perhaps you have Frank the fiddle-leaf or Phil the philodendron living in your bedroom? Or maybe you're just looking for a way to remember the botanical names of your plant babies. Yes, you could keep the tag

that comes with the plant, but sometimes plants don't have tags or you tend to "misplace" them. Plus, let's face it, those plastic tags aren't all that attractive. This is a fun way to remember the name of your plant, and it includes using something that might otherwise be thrown away. In this project, you'll upcycle table knives into decorative plant

stakes. Thrift stores usually have bins of inexpensive silverware, so you can buy multiples. When choosing your knife, look for the largest handle you can find so there's more room to decorate. Then the fun begins! This is a great project to create with a group. Share paint and decorations with friends and have a fun evening crafting for your plants.

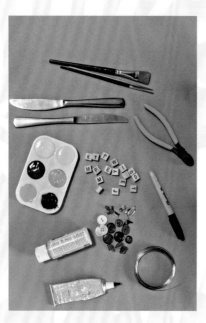

MATERIALS

Table knives
(not sharp knives)

Acrylic paint*

Paintbrushes
(assorted sizes)

Outdoor sealer (optional)

Something to stick your
knives in so they can dry
in an upright position
(I used a piece of foam,
but you could use a pot full
of soil or a plant.)

Covering for the surface you
are working on, such as a
drop cloth or newspaper

Disposable gloves (optional)

Small wooden alphabet tiles
(I used ½" [1 cm] tiles.)

Industrial-strength adhesive

Wire (I used 20-gauge
beading wire.)

Assorted charms or buttons

Wire cutters

Permanent black marker
(optional)

*Note: Do not get the paint on your clothes. It will stain. If working
outside, you can use spray paint specifically formulated for metal.
If indoors, use a liquid form of acrylic paint.

GETTING FANCY . . . OR NOT

This project can be as involved or as
simple as you choose. Leave the knife its
original silver color and simply glue on
letter tiles to make the word you'd like, or
paint the handle and handwrite the words.
Use wire, small charms, and beads to
make more decorative knives and consider
them jewelry for your plants!

STEP 1 Paint your knife handle and allow it to dry, placing it in an upright position. If you're applying outdoor sealer, do so at this time and let it dry. (Note: The outdoor sealer is not necessary, but it does help keep the paint from chipping off.)

STEP 2 Place your alphabet tiles on the handle first to ensure spacing and fit. Place a small amount of glue on each tile and attach it to the knife. Leave the knife lying flat until the glue is dry. This keeps the tiles from shifting or sliding off. (Note: Use caution when using the glue. Do not get it on your hands, clothes, or furniture.)

STEP 3 Cut a piece of wire approximately 24" (61 cm) long. Select a button or bead and string the wire through it until it is in the middle of the wire. Put a drop of glue on the button or bead and

attach it to the handle of the knife right where the blade ends and the handle begins. Bring both ends of the wire to the back of the handle and twist it to keep the button in place. Let the glue dry. Now you are ready to crisscross the wire up the knife handle. Cross the ends over the front of the knife, bring them to the back again, and twist to secure. Bring the wire back to the front and crisscross it again. If you have written a word or glued tiles on, make sure your crosses occur between the letters. Twist the wire again on the back, and bend the leftover ends of the wire out to either side of the knife.

STEP 4 Twist one side of the extra wire around the handle of a small paintbrush to make a loop, add the charm or bead, and squeeze the loop together to secure the charm.

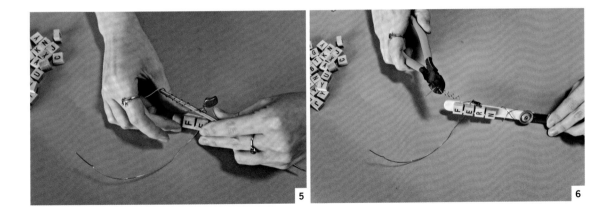

STEP 5 Twist the excess wire around the paintbrush handle to make a spring-like decoration, and cut off the excess wire. (Twisting the wire around the paintbrush handle is for decorative purposes only; you could choose to attach the charm or bead and then cut off the excess wire.)

STEP 6 Repeat the directions above on the wire from the opposite side, adding another bead or charm if you wish.

The fun of making these plant stakes is that you can use your imagination and creativity to make them your own, combining the paint, tiles, wire, and charms, or keeping them simple.

A PAINT-ONLY OPTION

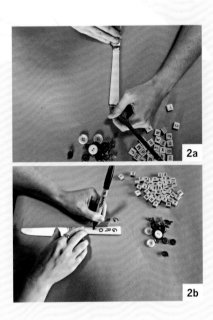

STEP 1 If you'd like to handwrite your own words on the plant stake, paint the knife handle a light color so your words will show up prominently. Let the paint dry completely and then handwrite your words with a permanent black marker.

STEP 2 Draw an outline of the knife handle on paper and practice writing your word to make sure it fits before writing on the actual handle.

STEP 3 After writing your words, apply the outdoor sealer, let it dry completely, and your stake is finished. If desired, add the wire and charms following the directions above.

WOODEN PLANT TRELLIS

Do you have a plant with long tendrils you don't know what to do with? Are you tired of stepping on it or watching your cat bat at it? You could make a macramé hanger (see page 48) and suspend it from the ceiling. Or, you could

use it to make a living wall. This project uses wooden dowels to build a trellis your plant will love scrambling through. It can be hung on the wall or even the ceiling. By using removable adhesive hooks, it's possible to hang it anywhere and then remove it easily (carefully taking the plant off it

first, of course!). This indoor trellis could also be inserted into a large container rather than being hung on a wall. If that's how you'd like to use your trellis, use wooden dowels made of cedar that can better withstand the constant moisture in the soil without rotting.

MATERIALS

8 dowels 3' (91.5 cm) in length (I used ⅜" x 36" [9.5 mm x 91.5 cm] square dowels purchased at my local hardware store.)

Small nails (I used #18 x ⅝" [1.6 cm] wire nails.) Note: Use nails that are no longer than the thickness of the two dowels you are nailing together, or you'll accidentally attach your trellis to the floor or work table.

Hand saw

Sandpaper (medium or fine grit)

Hammer

Tape measure

Painter's tape

Paint or stain (optional)

String or twine (optional)

Industrial adhesive or wood glue (optional)

Vining plant such as pothos, English ivy, or grape ivy

Removable adhesive hooks or nails

WHERE TO BUILD

You'll be hammering nails into wood, so construct this project on the floor or a sturdy table. Protect your work surface with a drop cloth or another material.

STEP 1 If the ends of your dowels are color-coded by size, use a piece of sandpaper to remove as much of the paint as possible.

STEP 2 Cut the five horizontal cross bars 21⅛" (54 cm) long. This length allows for a 5" (13 cm) overhang on each side and accounts for the ⅜" (9.5 mm) width of each vertical dowel.

STEP 3 Lay three 36" (91.5 cm) dowels parallel to each other and 5" (13 cm) apart on your work surface. Make sure they are even at the top. Use small pieces of painter's tape to firmly attach the three dowels to the surface you're working on. This is a big help, especially when attaching the first horizontal dowel.

STEP 4 Lay the first horizontal 21⅛" (54 cm) dowel across the three vertical dowels 5" (13 cm) down from the top. The horizontal dowel will extend out 5" (13 cm) on each side of the vertical dowels. I used a small amount of glue at each spot where the dowels intersect for added stability. It is not crucial but helpful.

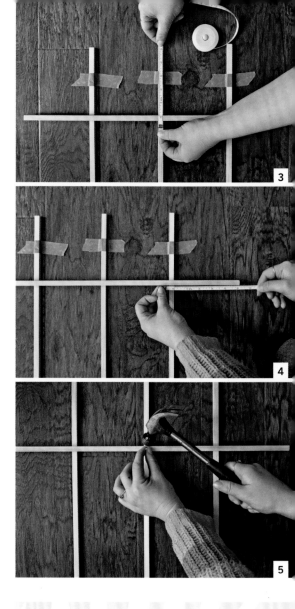

ADDING A SPLASH OF COLOR

You can make your trellis colorful with paint or stain, or you may choose to leave it natural and let the plant be the star of the show. I wrapped colorful twine around each intersection to add a small touch of color to my trellis.

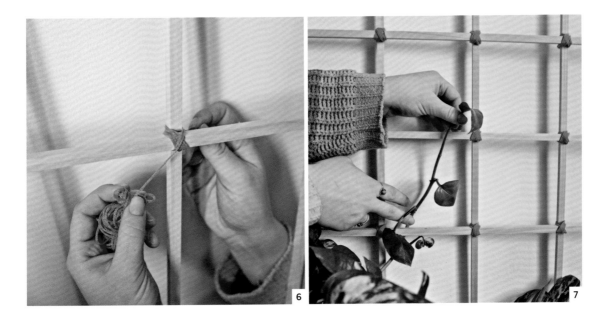

STEP 5 Nail the dowels together in the three places where they cross each other. Make sure to center the nail as close as possible to the middle of the two intersecting dowels. Otherwise, the nail may come out the side or split the wood. Continue adding the remaining four dowels in 5" (13 cm) increments down the trellis's length.

STEP 6 Use colored twine to cover the nails and give the trellis the appearance of being tied together at the junctions. The twine isn't necessary, but it adds some color to the trellis without taking too much attention from the plant. Use different colors of twine on every intersection to make it really unique.

STEP 7 Using the removable adhesive hooks, hang your trellis on the wall. Carefully weave the long plant tendrils through the openings, trying not to bend or break them. If you have a plant on a shelf, hang the trellis from the ceiling. No matter where you place your trellis or how you decorate it, I guarantee your plant will love climbing through it!

MORE OPTIONS

Taking into consideration the width of the dowels as you nail them together, there should be approximately 9½" (24 cm) left at the bottom of the trellis. You could cut 4½" (11 cm) off the vertical dowels before assembling if you'd like the trellis to be a perfect square.

This trellis can be made any way you like. The only requirement is it must have openings for the plant to climb through. You could place your dowels in an abstract pattern, nailing them together wherever they cross, if you wish.

SWINGING PLANT HANGER

Have you been looking for a different kind of plant hanger to mix in with your macramé? This swinging hanger is a simple project to

make and a unique way to show off your plants. The juxtaposition of the wood, metal, and string is unique. You can create this hanger in less than an hour, and your plant can be displayed in no time. This is the perfect perch for an air plant or for a plant your kitty has been batting at.

MATERIALS

2 8" (20 cm) metal rings

3½" to 4" (9 to 10 cm) wood round (A more oval-shaped piece of wood works best.)

Wooden large-hole macramé bead

3 yards (2.7 m) cotton macramé cord or twine

4 small cup hooks (only need to be large enough to accommodate the metal rings)

Marker

Ceiling hook

Air plant or small potted plant

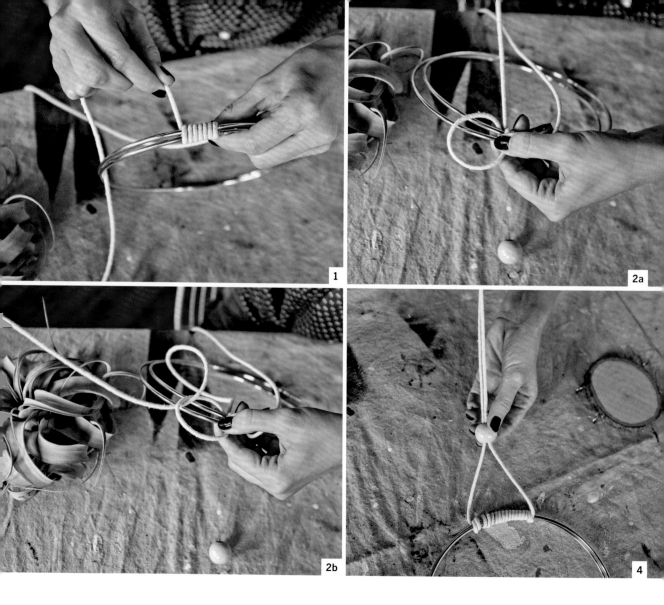

STEP 1 Place the two rings together on your work surface. Fold the 3-yard (2.7-m) cord in half and bring one end through both rings, stopping with the center of the cord at the top of the rings. The two ends of the cord will be used separately. Using one end of the cord, wrap the cord tightly around the two rings for 1½" (4 cm) along the rings.

STEP 2 Leaving the last wrap loose, bring the end of the cord around the rings, thread it through the loop, and pull tightly. This keeps the cord from loosening up on the rings.

STEP 3 Using the other end of the cord, wrap the cord around the rings 1½" (4 cm) the opposite way, again keeping the last wrap loose, threading the end of the cord through as before and pulling tightly. You now have 3" (7.5 cm) of wrapped cord across the top of the rings with the loose ends of the cord on each end.

STEP 4 Bring the two cords together and, using the large macramé bead, thread the bead over the two cord ends. Using the two cords, tie a knot at the level where you would like it to hang. Cut off

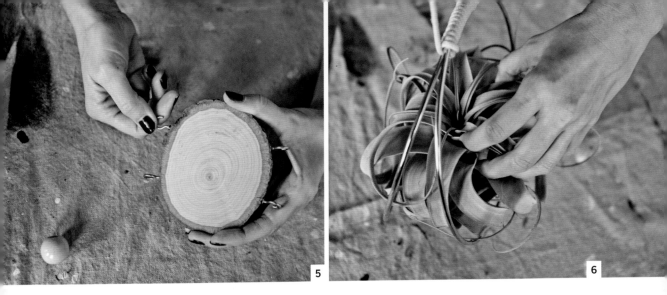

5

6

the excess cord or leave it to hang, in case you need to change the height of the hanger later. (If you would like a much longer hanger to accommodate a loft with tall ceilings, just use a longer piece of cord from the beginning.)

STEP 5 Now we are ready to add the wood round to the bottom of the rings. It works best if you first hold the wood round between the two rings to see where it is best to place the four hooks, as every piece of wood will be a different shape. The cup hooks need to be able to hook over the rings, so choose the place to attach them carefully. You may want to mark the spots with a marker. Screw the hooks into the side of the wood round, making sure all the hook openings are pointed down.

STEP 6 Place the cup hooks over the rings, and you're done!

MORE IDEAS

If you want to use larger rings to make a swing for a bigger container, double the rings on each side for added strength. Also, find cup hooks that are heavy duty and will be able to support the weight of the plant you are placing on the hanger. You don't want the plant to bend the hooks and fall, injuring the plant and anything that may be under the hanger.

This swinging plant shelf can be hung against a wall or from a ceiling hook. It's perfect for an air plant or a small potted plant. The most important point is to choose a plant that will work in the light levels you have to offer it.

This is an easy, quick project to make, giving your plant a place to "hang" out and swing in the sun.

KOKEDAMA

String gardens have become popular on social media, but if you haven't seen one yet, a string garden is a grouping of kokedama balls suspended from string to make a garden. In Japanese, *kokedama* means "moss ball." It is simply a plant that has its roots wrapped in moss instead of being grown in a traditional container.

Kokedama can be displayed by being hung or placed in a shallow dish. It's an interesting and unique way to display houseplants. Any plant can be used for kokedama, but make sure the plant you want to use will grow well where you want to display it. Because the root system is only contained in moss, it will dry out more quickly than it would in a container. You'll have to water your kokedama more often than your plants in containers, so using a plant in high light, where the plant would require more water, may not be the best choice. If you are someone who likes to water, go for it. Otherwise, choose a medium-

to low-light plant to place in a lower light level so you won't have to water as often. I hang a trio of these on my front porch every summer. They receive morning sun and are protected from the wind, so I only have to water them every few days. I take them down and soak them in a large pan of water until I know they have become completely moistened. They can drip on my porch, but when you water them inside, they'll have to be left in the sink or shower until they're done dripping. This is a project that is fun to do with friends. Gather all the supplies, have them bring a plant of their choosing, and make an evening of it!

MATERIALS

Green sheet moss
4" to 6" (10 to 15 cm) plant
Waxed cotton thread or colorful twine
Potting soil or bagged topsoil
Rubber gloves
Newspaper or plastic sheeting

STEP 1 Lay down newspapers or plastic to protect your work surface from mud and water. Disposable gloves are a good idea to protect your hands. Soak your sheet moss in a bowl of water to make it easier to work with. Add water to the soil you are working with too, so that it holds together when you form the soil ball to wrap the roots with.

STEP 2 Take your plant out of its pot and remove most of the potting medium from around the roots, being careful not to damage them or the plant.

STEP 3 Make a firm ball of heavy potting soil (I use bagged topsoil) that doesn't fall apart. Many kokedama sources call for something called akadama soil. This is a clay soil that holds water and helps hold the ball shape. It's often used in bonsai soils. Akadama isn't readily available and is a bit costly. I have been making kokedama balls for years and have never had a problem with using topsoil.

STEP 4 Squeeze the excess water out of the moss and lay out a circle of wet sheet moss large enough to cover the root ball. The green part of the moss

THE BEST THREAD

Using twine or cotton thread that isn't waxed or acrylic isn't a good idea. Because your moss ball is wet most of the time, cotton or jute twine will rot much faster. Trust me, if your kokedama is hanging, it will end up on the floor. What a mess! Waxed or acrylic twine lasts for a long time. If you choose to display your ball in a shallow dish or on a stand, the thread choice isn't as crucial.

KOKEDAMA CARE

Check your kokedama ball often. It may dry out faster than your other plants. Make sure it is kept moist. Soak it in the sink, submerging the ball in water, and leave it there until the ball is completely moistened. Then, allow it to drain and hang it back up or return it to its bowl.

Make one or make multiples to hang in a string garden in front of your window. Kokedama balls add a unique touch to your indoor garden and will mix well with your other hanging plants.

will be facing down so that when the root ball is wrapped, the green side will be showing.

STEP 5 Split the soil ball in half and add one half to each side of the root ball of your chosen plant.

STEP 6 Gently squeeze it back together, forming a sphere.

STEP 7 Set your plant in the center of the moss and bring the sheet moss up around the root ball. Make sure the entire root ball is covered, but not the stems of the plant.

STEP 8 Wrap your thread around the middle of the ball and tie a knot to secure it. Tuck the short end in as you start wrapping the long end. Wrap the ball, turning it as you go to make sure the ball is sufficiently covered. The twine should be wrapped snugly around the ball so that it doesn't become too loose when the ball dries out, but not tight enough to damage the plant roots.

STEP 9 If you'd like to hang your kokedama ball, cut three or four equal-length pieces of cord. Cut the cord long enough to be able to hang your kokedama at the height you would like it. Knot each string to the wrapping cords at different spots around the ball so that it hangs evenly. Knot the strings together toward the top, and you are ready to hang your kokedama.

GRAFTING CACTI

A grafted cactus consists of two different cacti fused or grafted together to grow as one. Did you know that if you use a fast-growing cactus and graft a slow-growing cactus on top of it, the slow grower will grow faster? Isn't nature amazing? Also, by grafting two cacti together you can make a new unique plant—

a form of botanical art. The process isn't complicated but does require a high level of preciseness and cleanliness to make sure the cacti graft together well. An important point is that you must use two cacti, not a succulent and a cactus.

It is easy to mistake many succulent plants (such as euphorbias, for example) for cacti because they have spines, but they won't graft to a true cactus, so be aware of what you are purchasing for your grafted masterpiece.

GRAFTING SAFETY AND CLEANLINESS

Because you'll be cutting plants and using a very sharp instrument, there are a few important points to address before you start this project.

- This process is quite simple, but make sure you are careful when using a razor knife. Always cut away from your hands and body in case you slip. Hold the cacti with the tongs to prevent the spines from puncturing your skin. This is a perfect project to do with a friend as it's much easier with a second set of hands.

- Do not use pruners or scissors. A knife or razor is best, because the plant needs to be cut with a back-and-forth motion for a clean, precise cut. Using pruners or scissors would pinch the end of the cut cactus.

- Don't leave the "wound" or cut area open to the air for longer than a minute, and make sure you don't touch the cut area or let it touch the surface you are working on. This would contaminate the surface of the cut. If that happens, cut a thin slice off the cactus again. Keeping the cuts clean and dirt-free is imperative. And between cuts, the knife should be sprayed or wiped with rubbing alcohol. Just make sure the excess alcohol is wiped off.

Now that we have those important points out of the way, let's get started!

MATERIALS

2 cacti, one columnar and one round—monstrose or crested cacti work well

A sharp knife (I used a razor knife)

Alcohol for sterilization

Tongs to hold cacti while cutting

Rubber bands

Gloves (optional)

STEP 1 Choose two cacti that you think would look good grafted together.

STEP 2 Carefully slice the top off the columnar cactus. This is called the stock plant. The stock plant keeps its roots and supports the plant being grafted to it. There is no exact place that this needs to be done. You decide how tall or short you want your grafted cactus to be. The stock plant cactus will only grow in girth; it will not grow taller, because the growing tip has been removed.

STEP 3 The skin of the cactus needs to be cut around the edge, beveling it slightly. This needs to be done because the soft center of the cactus may sink in, which would leave the outside (skin) taller than the center, preventing the two cacti from attaching to each other; they need a good connec-

tion to graft together. The beveled edge allows the center to sink a bit and yet still have good connection with the cactus above. (Note: In the photos, I exaggerated the amount of edge to cut off, for clarity purposes. Only a small amount needs to be taken off.)

STEP 4 Before making your next cut, spray or wipe your knife with alcohol.

STEP 5 Slice the round, crested, or montrose cactus off the root ball and immediately place it on the cut top of the columnar cactus, making sure to match up the growing points **(5a)**. These should appear as round outlined areas in the center of the cut area. The top plant being grafted onto the bottom stock plant is called the scion **(5b)**.

4

7

STEP 6 Carefully press the scion down to get a firm connection between the two cacti. Use rubber bands to hold the two cacti together. This is where a friend's help comes in handy. Stretch the rubber bands around the pot and cacti while holding them together. If you are doing it alone, tape the rubber bands to the bottom and sides of the container. Then it is easy to grab the bands and bring them up around the top of the cacti without them coming off. This can be challenging, especially with spines.

STEP 7 If there is any shrinkage of the cacti as the cut ends callus over, the rubber band will contract with the plants, keeping them in tight contact with each other. Now it's all about patience.

AFTER-GRAFTING CARE

The newly grafted plant needs to be in bright light to full sun, and it shouldn't be moved excessively. The two cacti need to fuse together, and moving them too much could break the connection.

It won't take long for them to fuse together and start growing. Within 4 to 6 weeks, the stock and scion will be growing as one.

BONSAI PLANTING AND PRUNING

Bonsai is an age-old Japanese art form that involves growing plants in small containers and pruning and training them to look like they

are old, full-grown trees. Often the plants used are hardy trees that require a winter rest—such as maples or beeches—and take many, many years to train. If, on the other hand, you choose a houseplant to train as a bonsai, you can achieve results much quicker. Let's go through the process step by step, from buying the plant to its long-term care.

MATERIALS

Suitable houseplant, preferably in a 4", 6", or 8" (10, 15, or 20 cm) container. I like using ficus or *Schefflera arboricola* (dwarf umbrella tree). Look for a trunk that has some character.

Bonsai soil

Aluminum bonsai wire, 2 to 2.5 gauge

Sharp pruners or scissors

Bonsai pot in the appropriate size for your plant

7-hole mesh plastic canvas

Chopstick (pencil will work)

2

3a

3b

4a

STEP 1 Study your plant and decide which side you would like to be the front. Find the side that is "greeting" you. Maybe it leans a bit, or the branches are reaching out toward the front. Every plant is different. You also need to decide which shape you'd like to eventually end up with, because your plant must be trained in this shape from the beginning.

STEP 2 After you have studied your plant and decided which form you want your tree to become, it's time to trim your tree. A good rule of thumb is to take off any branches that point straight up or straight down. If the trunk of your plant is thin, a shorter tree makes the trunk look larger. The fun part of this process is that it's all up to you. You decide what your tree will eventually look like by the way it is trimmed and wired.

STEP 3 That leads us to the next step: wiring the tree. This is when you'll determine the final form of the tree. Starting with the bottom branches, cut your wire long enough to go from the tip of one branch, around the main trunk, and up the opposite branch. At this time, you can remove all the leaves to see the branches better. This will also help the leaves to regrow in a smaller size, more proportionate to the size of the tree. Usually there are branches across from one another, and you can use one piece of wire to train both branches simultaneously. By going around the trunk, there's more stability. Wrap your branches carefully, making sure the wire isn't too tight, but firm enough to hold the branch the way you'd like it to grow. Continue wiring the branches until you reach the top of the tree. If the branches are thin on top, you may have to use a smaller wire than you used on the bottom branches. Look down at the tree from the top. There should be no branches over top of lower branches. It should look like the spokes of a wheel. Check the wire often to make sure the plant isn't being scarred by a too-tight wire. Remove the wire just before it starts cutting into the branch, usually after a few months. If your branch moves back, it needs to be rewired.

STEP 4 Get your pot ready for the tree. Cut two small squares of the 7-hole mesh and cover the drainage holes in the bottom of the pot. This ensures that the bonsai soil doesn't wash out of the holes. Cut a 10" to 12" (25 to 30 cm) piece of wire and thread it through the drainage holes so

4b

5

6

that the ends of the wire come up into the pot. This wire will be used to secure the plant into the pot.

STEP 5 Now it's time to take the plant out of its nursery pot and get it ready to transition to its bonsai pot. This process may seem harsh, but it won't hurt the plant. Remove the plant from its pot and remove all the soil from the roots. A chopstick works well to tease the roots from the soil. Rinse the roots in water to remove any remaining potting soil. Because it's going into a much shallower pot than the one it came out of, the roots will need to be trimmed. Trim them just enough to fit into the pot. Position the plant the way you want it, then cross two ends of the wire over the root ball and twist the ends together to hold the plant in the pot.

STEP 6 Once your bonsai is in place, add bonsai soil to the pot, filling in between all the roots. Use the chopstick to work the bonsai soil down into the roots. Water it well.

CARING FOR YOUR BONSAI

Place your finished bonsai in a bright window, such as an eastern exposure. Because the pot is shallow, this special tree needs water more often than a regular houseplant. The rule is to check the pot every day, but not necessarily water it every day. When the soil mix has turned a light color and is lightweight, the plant probably needs a drink. Make sure to water the entire root ball to evenly distribute the water to all the roots.

For the long term, your plant will need to be trimmed regularly; the frequency depends on the plant. It will need to be trimmed at least once a year. Fertilize your bonsai once a month while it's actively growing in the spring and summer. Use a balanced fertilizer, such as 20-20-20. The plant will normally need to be root-pruned and repotted with fresh soil every 3 to 4 years.

SHADOW BOX TRIO FOR AIR PLANTS

This trio of shadow box shelves is a perfect accent for any wall in your bedroom, living room, or bathroom. With a few simple steps

and the right tools and supplies, you can make these shadow boxes in a short amount of time. The boxes have twine strung randomly throughout the frame, making places for air plants (tillandsias)

to hang out. You can paint the boxes to match your décor or stain them. Use your imagination and have fun decorating your shadow boxes!

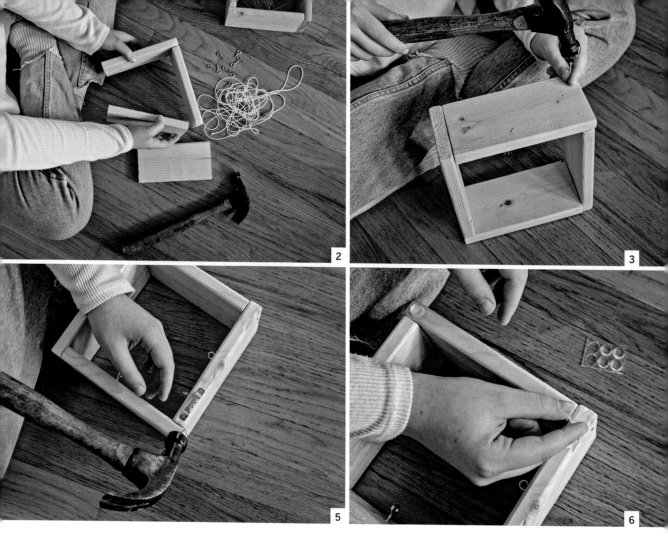

STEP 1 Using the square, draw straight cutting lines across the 1" x 4" x 8' (2.5 cm x 10 cm x 2.4 m) board in 7-inch (18-cm) increments. Using a square ensures your lines are straight when you draw the cutting lines. If the lines aren't straight, your box will not be square. For three boxes, cut twelve boards of the same length.

STEP 2 Start by nailing one board to the end of another board, with one board overlapping the other. Use at least two nails on each end. Turn clockwise so your two boards form a backwards L shape.

STEP 3 Nail the next board on top of the backwards L to form a sideways U, again overlapping the two boards. Turn the sideways U clockwise so it looks like a U, and again nail a board on top of the right

arm of the U. After nailing the final edge together, you now have a box. Repeat the directions above twice more to create two more boxes.

STEP 4 Paint or stain your boxes the color of your choice, or leave them natural.

STEP 5 Fasten two sawtooth picture hangers to the back edge of the top board of each box, placing one on each end. If you would like to hang your shelves in a diamond pattern, attach one hanger diagonally across a corner of the box.

STEP 6 Stick two rubber bumper pads on the bottom back edge of the box, one on each end to ensure the box is level against the wall. These bumpers protect the wall, but they also compensate for the depth of the picture hangers. There will be a slight space between the box and the wall because of the hangers and bumper pads. Use a level when hanging the boxes if you want them hung straight.

STEP 7 To make a web inside the box for your air plants, screw three eye screws randomly into each interior side of the box.

STEP 8 Tie your twine to the first eye screw and then string it randomly between the other eleven eye screws to make a web-like design, knotting it to the last hook. Trim the ends of the twine.

STEP 9 Place your air plants in the openings created by the twine.

PLACEMENT OF THE BOXES

Your air plant shadow boxes need to be hung where they will receive some bright light from a window. Hanging your boxes next to the window on the same wall might not be the best place for them, because they will never get any direct light. Instead, hang them on the wall across from the window (if it isn't too far away) or on the neighboring wall so when the sun shines in, it gives the plants some light. If you don't have enough light from the windows, a small clip-on LED light is the perfect solution. Remember, your air plants have to be removed and soaked often. Enjoy your living work of wall art!

PLANT SHELVES

These shelves are fun to make and even more fun to decorate. You may have a collection of small items that would be perfect on these shelves, side by side with your plants. Let your personality show, using tiny gnomes, gems, an owl collection,

or whatever. These box shelves are made in three sizes; they can be hung together in a pleasing arrangement or on different walls individually. You could also make them all the same size, if you'd like. When you're finished building the shelves, paint or stain them, or leave them natural. It's all

up to you! When decorating them, choose plants that work with the amount of light you have to offer. If the light you have isn't enough for your plants, purchase a small clip-on or hanging grow light to add light to your shelf.

MATERIALS

12 feet (3.7 m) of 1" x 4" (2.5 cm x 10 cm) wood (I used cedar.)

Hand saw (A power saw works better if you have one.)

Nails (I used wire nails #16 x 1½" [3.8 cm].)

Hammer

Paint or stain (optional)

Pencil

Square

6 1¾" (4.5 cm) sawtooth picture hangers

6 clear rubber bumper pads

Level

Sandpaper in different grits (coarse, medium, and fine)

Small plants and collectibles

Small clip-on LED grow light (optional)

PAINT, STAIN, OR NATURAL

Paint or stain your plant shelves any
color you choose, or leave them natural.
If you decide to paint or stain, allow them
to dry completely before hanging and
adding plants.

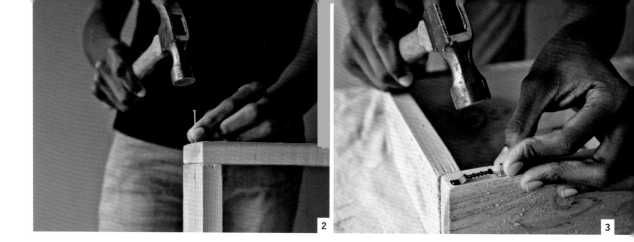

Cutting the boards:

STEP 1 Using the square, draw straight lines (cutting lines) across the 1" x 4" (2.5 cm x 10 cm) boards as follows:

For the large box shelf:	*For the medium box shelf:*	*For the small box shelf:*
2 12" (30 cm)	2 10" (25 cm)	2 8" (20 cm)
2 10¼" (26 cm)	2 8¼" (21 cm)	2 6¼" (16 cm)

IMPORTANT TIPS FOR BOARD CUTTING

The three shelves are each a different size, so be careful when cutting the boards. "Measure twice, cut once," is an excellent mantra to keep in mind. Unlike the air plant frames (page 86), the sides are different sizes for each of these shelves. Each one has two long sides and two shorter sides. The two long boards will be on the top and bottom, and the two shorter boards will be the sides of the shelf.

Using the square ensures that your lines are straight when you draw them. If the lines are not straight, the board will not be straight when it is cut. This would lead to an odd-shaped shelf, not a square. Using the sandpaper, sand the boards to remove any rough edges. Start with the coarse grit, switch to the medium, and then use the fine grit for a smooth, no-sliver finish.

Assembling the shelves:

STEP 2 Using the two 12" (30 cm) and 10¼" (26 cm) pieces and two nails per end, attach the two smaller pieces to the 12" (30 cm) piece to form a U shape. Make sure the 12" (30 cm) board is placed on top of the two 10¼" (26 cm) boards. Then attach the remaining 12" (30 cm) piece to the bottom of the two 10¼"(26 cm) pieces to form a square. Repeat the process using the two 10" (25 cm) and 8¼" (21 cm) pieces, and the two 8" (20 cm) and 6¼" (16 cm) pieces. You now have three square box shelves.

4

Finishing the shelves:

STEP 3 Fasten two 1¾" (4.5 cm) sawtooth picture hangers to the top edge of the top board on the back of the shelf, placing one on each end.

STEP 4 Stick two rubber bumper pads to the bottom edge of the bottom board on the back of the shelf. This ensures the shelf will sit level against the wall. The bumpers protect the wall and compensate for the depth of the picture hangers. These shelves are perfect for adding some greenery to any room and displaying a few of your favorite things.

PLACEMENT OF THE BOXES

Before hanging the shelves, consider the light in the room where you're hanging them. They need to be hung where they'll receive some bright light from the windows. If you hang your boxes right next to the windows on the same wall, they most likely won't receive any direct light. Instead, hang them on the wall across from the window (if it isn't too far away) or on the neighboring wall where some sun may reach them. Hang the shelves in any pattern you choose, using a level when hanging them so that items won't slide off. If you find your plants aren't receiving enough light, you can purchase a small clip-on or hanging grow light that would boost the light your plants are receiving.

PICKING THE BEST PLANTS

Use plants in 2" to 4" (5 to 10 cm) pots so they fit easily on the shelf. Make sure to use saucers under your plants to catch any excess water so it doesn't run down the wall. You can also take your plants to the sink to water them, let them drain, and return them to the shelves. Check the plants often for water needs. Since they may only receive light from one direction, turn your plants often to make sure they grow symmetrically.

PROPAGATING PLANTS BY AIR LAYERING

If you have a plant so tall that it is hitting the ceiling, or one that has greenery only at the top, with bare, straggly stems beneath, then this is the project for you. Many people throw away overgrown or unsightly plants, not knowing they could easily save the plant with a simple propagating technique called air layering (or air

propagation). Instead of taking a cutting and rooting it in water or soil, this technique takes place directly on the plant, typically on the main stem or a side shoot. It's often used on plants with woody stems, such as citrus, fiddle-leaf fig, dracaena, and rubber plant, to name just a few. Many plants that we use as houseplants become large trees in their natural habitats and have bark-covered stems. This tropical bark isn't as thick as an oak

or a maple, but it's woodier than soft-stemmed houseplants such as aglaonema or dieffenbachia. While soft-stemmed plants are easy to propagate in moist potting medium or water, woody cuttings are tough to start in water or soil. Air layering works perfectly for these tougher-skinned plants. However, you'll need a bit of patience. It takes a few months for the roots to form.

MATERIALS

A plant that needs to be shortened or has a bare stem

Sphagnum moss (often sold as orchid moss)

Plastic food wrap

Sharp knife

Twine

Toothpick, matchstick, or bamboo skewer

Patience

An appropriately sized pot and potting medium for the newly rooted plant

Disposable gloves

PICKING THE RIGHT SPOT

Before you begin, determine where on the trunk of your plant you'd like to perform the air-layering process. If the tree is hitting the ceiling, you may want to only drop down 1 to 2 feet (30 to 60 cm) from the top. Where you choose to air layer determines the final height of the plant.

STEP 1 Soak your sphagnum moss in a bowl of water—it's best to work with moistened moss. A large handful of moss is all that's needed, and you may want to wear disposable gloves.

STEP 2 Use a sharp knife to carefully slice with an upward motion, approximately one-third to halfway through the stem of the plant at the chosen spot. Do this carefully, making sure to not cut yourself or cut all the way through the stem.

STEP 3 Wedge the toothpick or matchstick into the cut to keep it open. Trim the ends of the toothpick even with the edges of the plant stem.

STEP 4 Wrap the moistened moss around the cut area. The moss ball should be about the size of an orange or a bit larger when you're done, depending on the thickness of the stem being air layered.

WHAT CAN YOU DO WITH THE MOTHER PLANT?

If you air layer a single-stemmed plant, you can throw the mother plant away or cut it down to the soil line and see if it sends out new growth from the bottom. Plants have a will to live, and with a healthy root system still intact, quite often they will send out new shoots. Now you have two plants!

If your plant has multiple stems and you only layered one of them, you could give the mother plant to a friend with higher ceilings. A better idea is to invite friends over, and each person could air layer a stem with you. Have them write their name on the plastic wrap so they know which stem is theirs. Later, after the roots have formed, have another gathering where each person pots up their own plant. It could be an air-layering party! And then a few months later, a potting party! Don't feel bad if you have to discard the original plant, though. You saved what you could, and the plant will continue living in your home, just in a smaller form. You made a new life for your plant!

STEP 5 Use a piece of plastic wrap to enclose the moss completely. Tie the twine around the stem of the plant at the top and bottom of the plastic wrap to keep it closed and the moisture inside. Make sure the twine isn't too tight, which may damage the stem.

STEP 6 The key to this process is to make sure the moss stays moistened. Untie the top twine often to check the moss for moisture. If it feels dry, add a small amount of water and tie it back up. Keep checking the moss, and eventually you'll see roots developing. When the roots have filled the moss inside the plastic wrap, it's time to slice the newly rooted plant from the stem of the mother plant (see sidebar at left).

PLANTING YOUR NEW PLANT

Remove the plastic wrap but don't try to remove the moss because it may damage the roots. Cut the stem right below the newly formed roots, and pot it up in an appropriately sized container. Your container may be disproportionate to the size of the plant, but it's imperative to choose a container that fits the root ball with a bit of room to spare. If you use a container that is too large, yet proportionate to the size of the plant, the roots may rot from being surrounded by too much wet potting medium. Check the pot often, and when the roots have taken off and filled the pot, it's safe to up-pot to the next size pot.

TAKING LEAF AND STEM CUTTINGS

Swapping houseplants with friends is a lot of fun. Thankfully, you don't have to give up your favorite plant in order to get a new one to try. Instead, learn how to propagate your plant babies and you'll always have plenty to share with friends

and family. Propagation makes duplicates of your plants, and stem and leaf cuttings are among the easiest forms of plant propagation. It can be done with all sorts of houseplants, and it's a simple process. Some plants, such as succulents and African violets, are easy to propagate using just one leaf. Vining plants and those with long stems are easy to propagate from a stem cutting. Not only does the process make more plants,

but it also improves the look of the original plant, keeping it from becoming overgrown. It's easy, but it does take a little courage. You're going to have to snap leaves off, cut stems, and behead plants. I used to be afraid of doing these things too. I was concerned I was hurting the plant or might kill it altogether. Don't worry. I'll walk you through the steps and you will have more plants to share in no time flat.

MATERIALS

Plants to be propagated
or revitalized

Sharp knife or pruners

Potting soil

Containers

Patience

Instructions for propagating succulent leaves

STEP 1 Begin with a succulent, such as an echeveria (shown) or aeonium. "Snap" a few leaves off the bottom of the plant. To do this, carefully ease your finger in between the rows of leaves and apply pressure to the leaf close to the stem of the plant, pushing in a downward motion. It should snap right off. Another way to do the same thing is to lightly hold the leaf between your finger and thumb and move it to the right or left until it breaks off. The plant needs to be well watered and the leaves turgid (firm) for this to work. If the leaves are not firm, they may not snap off easily and you may damage the plant.

STEP 2 Fill a pot with moistened potting soil. It can be large or small, depending on how many leaves you have to propagate.

STEP 3 Lay the leaves on top of the moist (but not wet) potting mix and forget about them for a bit. Do not stick the end of the leaf into the medium because the "snapped" end needs to heal or callus over before it can be planted. If they are freshly cut and inserted into moist medium, they will rot. By simply laying them on top of the soil, the ends will heal, and they will send out roots when they are ready.

PATENTED PLANTS

If a plant is patented, it may not be reproduced through propagation and sold. The plant tag will indicate if it's a patented plant.

STEP 4 Roots will soon develop from the cut end of the leaves and grow down into the potting soil. A tiny new plant will appear at the base of each leaf in a few weeks.

STEP 5 When they're a good size, you can pot each little baby plant into its own pot. Carefully remove the mother leaf if it hasn't already rotted away. There is nothing cuter than a tiny succulent!

Instructions for stem cuttings

STEP 1 Cut off a 4" to 6" (10 to 15 cm)–long piece of vine or plant stem, making sure the piece has at least one set of leaves on it.

STEP 2 Insert the cut end of the stem into a clean pot of moist potting soil. You can dip the cut stem in plant rooting hormone prior to sticking it into the soil. This isn't necessary for most houseplants, but rooting hormone can improve your chances of success and speed the growth of new roots. Alternatively, insert the cut end of the stem cutting into a container of water.

STEP 3 If you're growing in soil, keep the cutting well-watered, but not sopping wet, and it should form roots in 4 to 6 weeks. Some of the original leaves may turn yellow and fall off. Don't worry; new shoots will form once roots are established. If you're rooting your cutting in water, when you see roots forming through the glass a few weeks later, you then can plant the rooted cutting into a container of potting mix.

PROPAGATING TOO-TALL SUCCULENTS

Without enough light, succulents often develop long, leggy, bare stems. There is a solution. You can use one leaf to make a new plant (see page 100), or you can cut the top of the plant off (behead it!) and use it to start a new plant. Allow the cut to callus over for a few days by leaving it sit on the kitchen counter or anywhere that's shielded from blasting sun (you can leave it longer with no problem). Then insert the bottom end of the cut-off plant into a container of moist potting mix. The good news is the bare stem left behind on the mother plant may also send out a new little plant(s). This can be removed later and potted up individually, or it can be left to grow as-is.

Instructions for leaf cuttings

REX OR RHIZOMATOUS BEGONIA: Remove the leaf from the petiole. Cut the leaf into a wedge shape with the main middle vein intact and stick the point of the wedge into moist potting medium. A new plant will grow in a few weeks.

CANE-TYPE BEGONIA: A vegetative node is needed for stem cutting. Place the stem cutting in moist potting medium. Keep it moist while the roots and baby plants are being formed.

AFRICAN VIOLET: All you need is one leaf with approximately 1" (2.5 cm) of the petiole left. Cut the leaf petiole on a slant to allow for more root

growing area and insert on an angle into a small container of moist potting medium. Use a stick (or pencil) to make a hole first so you don't damage the petiole when inserting it.

I cover the leaf with a clear deli container to maintain a high, consistent humidity level. This isn't necessary if the potting medium is kept moist. New plants (yes, more than one violet) should appear in a few weeks.

When your begonia and African violets are 1" or 2" (2.5 or 5 cm) high, transplant them into new containers.

ACKNOWLEDGMENTS

I have always been what many call a "crafty" person. I know how to knit, crochet, macramé, cross-stitch, embroider, and more, and I enjoy creating projects with all these mediums. So, I thought it would be easy to write a book with projects. It wasn't as easy as I thought. Writing clear, concise, understandable step-by-step instructions was challenging. But with the help of an amazing team of people, it became easy, fun, and a great learning experience.

I'm going to attempt to thank all the wonderful people who listened, gave me advice, and helped so much with this book. First, a huge shout out to Heather Saunders, whose expert advice on each and every project was immeasurable. Your input and photos were priceless! To her partner Robert Brusseau, whose help with building projects, hanging plants, backdrops, and shelves, and acting as design coordinator with Heather was amazing. Thank you also to Robert and Heather for opening their home and garage for photo shoots. Thank you to all the models in the book, Hayley Steinkopf Bonafede, Kelly Ardito, Joy Crocker, TJ Williams, Christina Quirk, Lily Stotz, Serena Pham, and Tara Boinpally. Thank you also to Harrison Saunders and Kelly Ardito for opening their home for a photo shoot, and to Kelly for her photo design assistance. Thank you again to Danielle Dirks for the use of her Detroit Airbnb loft for photos. To my brother Brian Eldred for his help making sure the wood projects made sense and the measurements were correct. To my sister-in-law Beth Eldred for her help with the macramé patterns, and to her and my brother Keith for taking such amazing care of my mother and giving me peace of mind. What a blessing to me and our entire family. To Kurt Smith of The Flower Market for his help with the bonsai project and for lending materials. To Michael D'Arcangelo of MD Botanicals and the members of the Michigan Cactus and Succulent Society for all their expertise with grafting cacti. To Rachel Nisch of Graye's Greenhouse for her help with the plants for the stem and leaf cuttings. To my best friend Jeanine Merritt for her editing advice, endless support, and listening ear. Thank you! To my close friends Colleen Burton, Julie Smith, Susan Martin, and Jean Mancos for their support and prayers, and to my church family as well. To my good friend Nancy Szerlag for her editing help and advice. To the team at Cool Springs Press who have been exceptional, especially Jessica Walliser, my acquisitions editor, who came up with many of the project ideas. She has been generous with her advice and was a true pleasure to work with!

As always, I give all the thanks to God, for through him all things are possible. And last, but never least, to my biggest cheerleader and support, my husband John. He is so understanding of the boxes of plants, projects, and stuff everywhere while I am writing (and when I'm not). I love you! Thank you to all my friends and family who have been here for me as I continue along this new path in life.

ABOUT THE AUTHOR

Lisa is The Houseplant Guru and features all things houseplants on her website www.thehouseplantguru.com. She is the author of *Houseplants: The Complete Guide to Choosing, Growing, and Caring for Indoor Plants*, and *Grow in the Dark: How to Choose and Care for Low-Light Houseplants*. She grew up in rural mid-Michigan immersed every day in nature. She spent many days bicycling to her grandma's house just down the road, where watching her tend to her houseplants, especially African violets, led to a passion for indoor plants.

Lisa has been featured in *Better Homes and Gardens*, *Real Simple*, and *First for Women* magazines. She has also written articles for HGTVgardens.com and is featured in the houseplant section of Allan Armitage's Greatest Perennials and Annuals app. Being an avid outdoor gardener as well, she writes a regular column for *Michigan Gardener* magazine. She lectures extensively around the country, spreading the word about the importance of houseplants and how to care for them, and has been interviewed online, in print, on radio, and for podcasts such as *On the Ledge*, *Bloom and Grow Radio*, and *Epic Gardening*.

Lisa is a member of numerous plant groups, including the Town and Country African Violet Society, the Michigan Cactus and Succulent Society, the Southeast Bromeliad Society, the Hardy Plant Society, and Garden Communicators. Plant societies are where likeminded people share their hands-on, personal plant knowledge with others. Join one!

Lisa cares for hundreds of houseplants in her home in the Detroit area, where she lives with her husband, John, and Henry, their cat. She spends as much time as she can with her two daughters, Hayley and Chelsea, and their families. She loves to visit conservatories and gardens during her travels, and she volunteers at the Belle Isle Anna Scripps Whitcomb Conservatory in Detroit. Lisa feels that every home, office, apartment, and rental space should have a living houseplant (or hundreds). There is a houseplant for every situation, whether with natural light or electric. Anyone can have a green thumb, because we all need a little green in our lives!

INDEX

Other Books by Lisa Eldred Steinkopf

Grow in the Dark:
How to Choose and Care for
Low-Light Houseplants
978-0-7603-6451-2

Houseplants:
The Complete Guide to
Choosing, Growing, and
Caring for Indoor Plants
978-1-59186-690-9

Houseplants:
A Guide to Choosing and
Caring for Indoor Plants
978-0-7603-6592-2